Picking & Weaving

Bijou Le Tord

FOUR WINDS PRESS
NEW YORK

LIBRARY OF CONGRESS CATALOGING IN PUBLICATION DATA

Le Tord, Bijou.
Picking and weaving.

SUMMARY: Focusing on the people and machines that make it happen, shows how cotton is grown, harvested, processed into fibers, dyed, woven into fabric, and sold for a variety of purposes.

1. Cotton—Juvenile literature. [1. Cotton. 2. Textile industry] I. Title.

TS1576.L47 677'.21 79-23457

ISBN 0-590-07642-6

PUBLISHED BY FOUR WINDS PRESS

A DIVISION OF SCHOLASTIC MAGAZINES, INC., NEW YORK, N.Y.

PRINTED IN THE UNITED STATES OF AMERICA

LIBRARY OF CONGRESS CATALOG CARD NUMBER: 79-23457

1 2 3 4 5 83 82 81 80

To the people of the mill

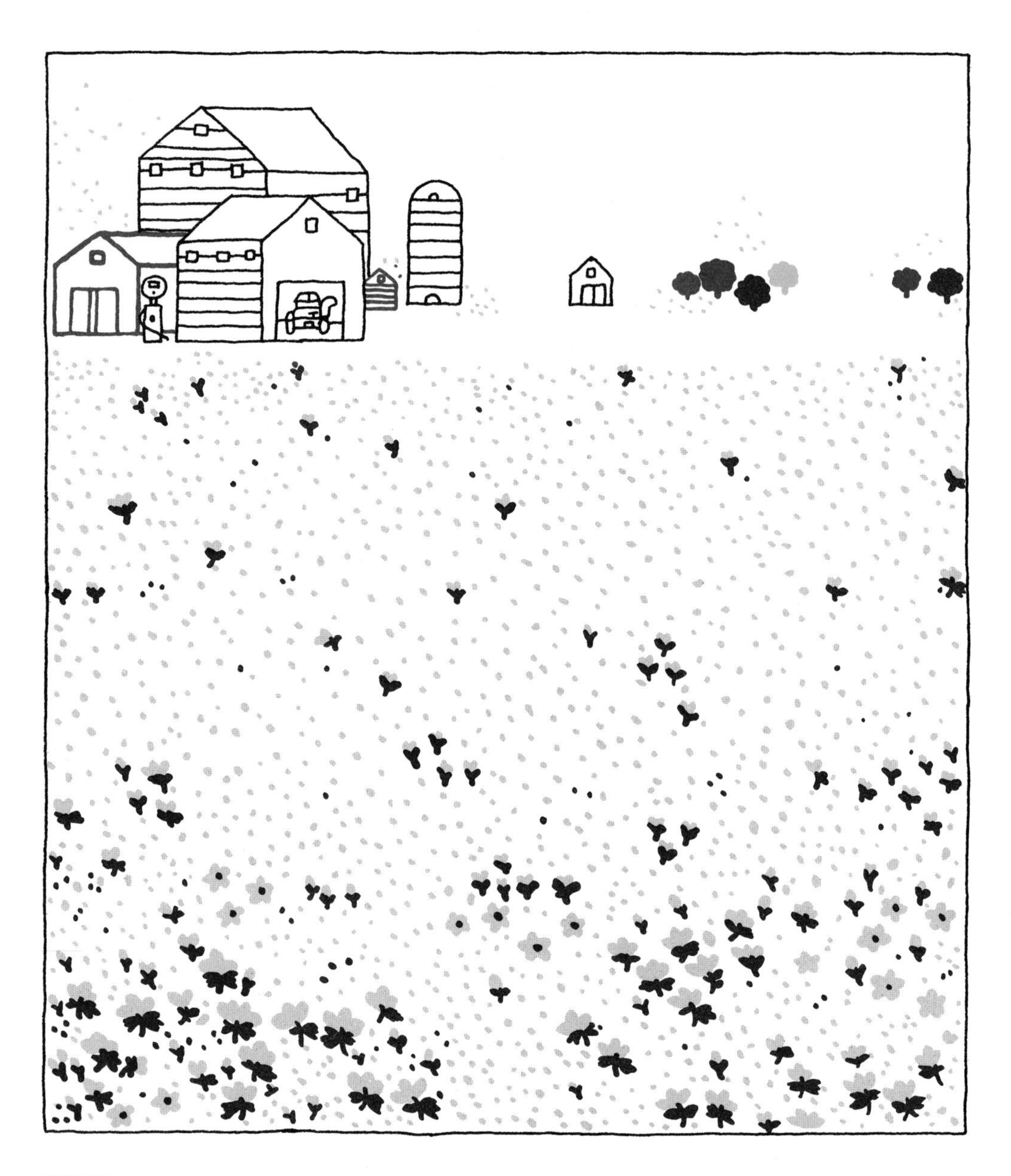

When the cotton plant flowers, its blossom is white or yellow. The next day, the flower turns red and falls to the ground.

Then a seedpod or cotton boll grows. A few weeks later, it bursts open showing the white, fluffy cotton fiber. Men and machines work all day, picking.

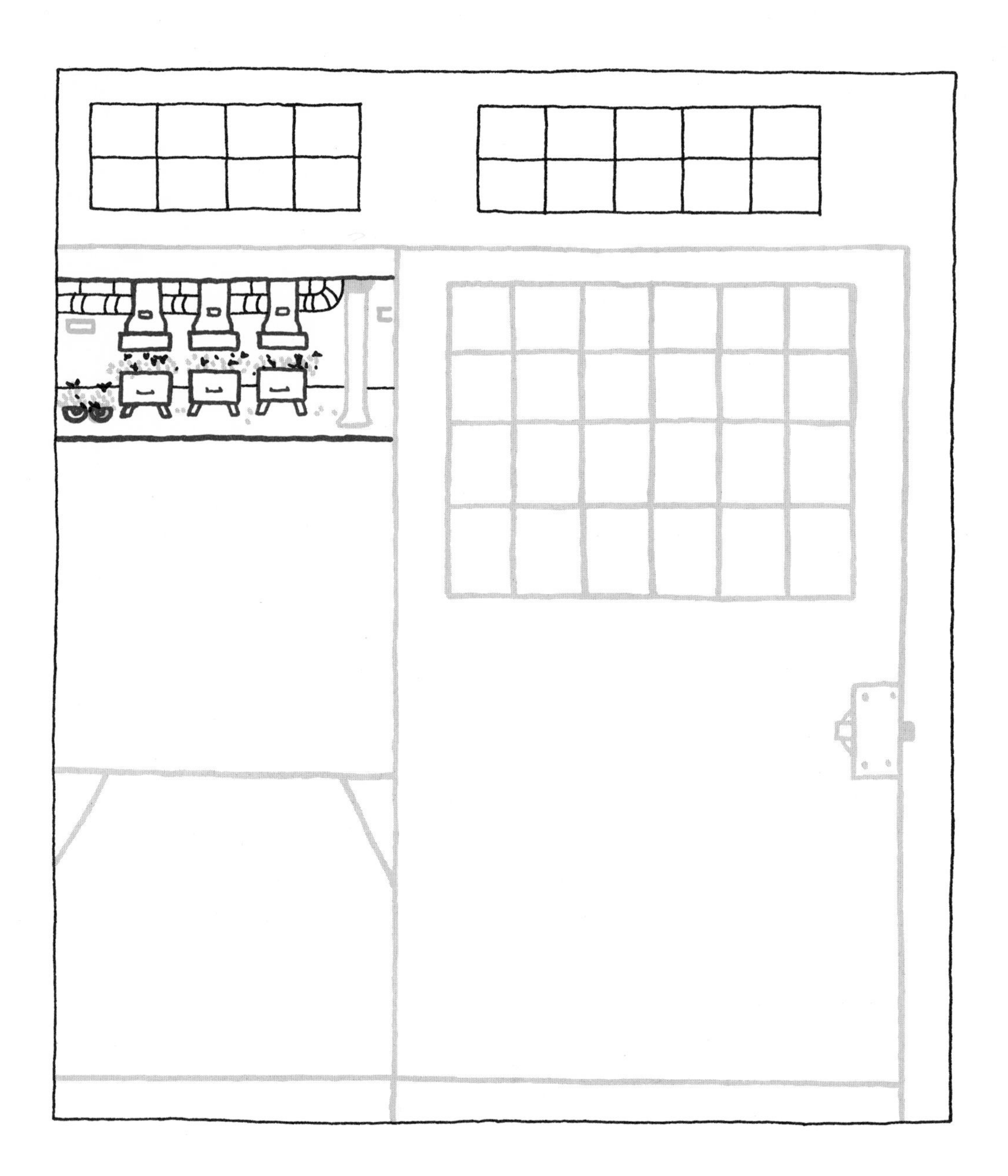

After harvesting, cotton goes to the ginning mill to be cleaned.

In the gin machine, seeds, small twigs, and dried leaves are separated from the cotton fibers. Then the clean cotton lint is sent to the textile mill.

Cotton is shipped in a high-rack truck. It is delivered to the mill tied in heavy bales.

After the cotton bales are unloaded and weighed, they are broken, fluffed, beaten, and rolled into a thick blanket called a lap.

The lap is fed to the carding machine, the giant of the mill! It cleans, untangles, and straightens the raw cotton fibers.

When cotton leaves the carding drum, it is gathered into a soft rope called a sliver. The sliver is neatly coiled in a tall can.

To make a strong, tight yarn, the sliver must be pulled and twisted through a series of rollers, which draw out the fiber into a finer and finer thread.

Finally cotton is spun into yarn in the spinning room.
Bobbins and spools spin, twist, and wind the yarn.

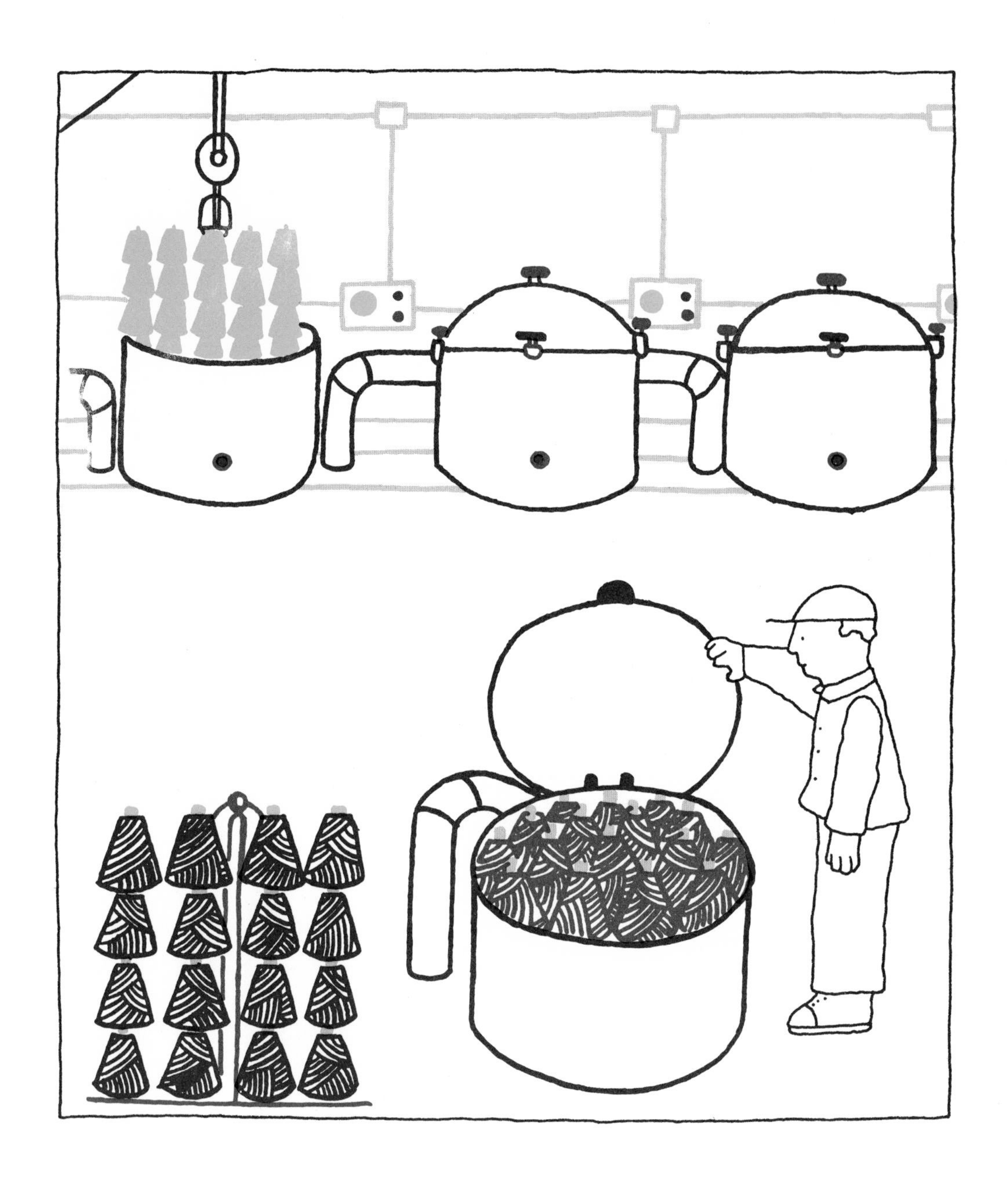

Big vats are filled with dye, and yarn cones are dipped in. Yarn-dyed colors are clear, clean, and bright.

Each spring and fall a textile designer creates plaids, checks, or striped patterns.
These designs will be woven into cloth.

At the mill, a technician reads a plan of the design from graph paper. She pulls in the warp ends. Later a pick filling is added on the mechanical loom. A pick filling is the yarn that crosses the warp threads horizontally.

One warp over, one warp under the filling,
one over, one under, one over, one under.
This is a plain weave.

A loom can weave plaids, tattersalls, stripes, checks, and other fancy goods.

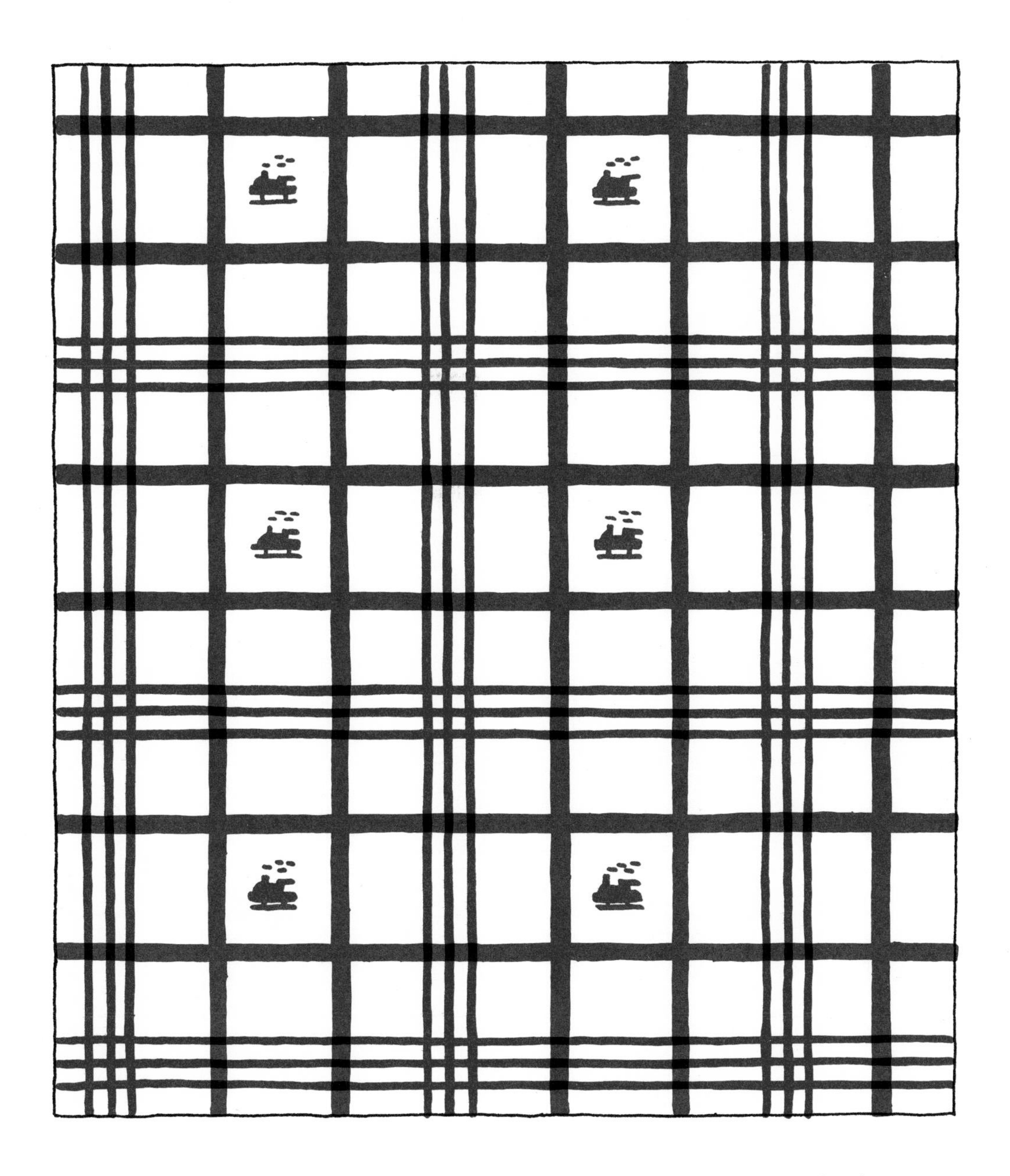

Sometimes, a small design or clipped-spot is added to the fabric, leaving a neat embroidered pattern.

Cotton, now woven, is washed, steamed, and dried. It is inspected for good quality weaving, and shipped to the storage room.

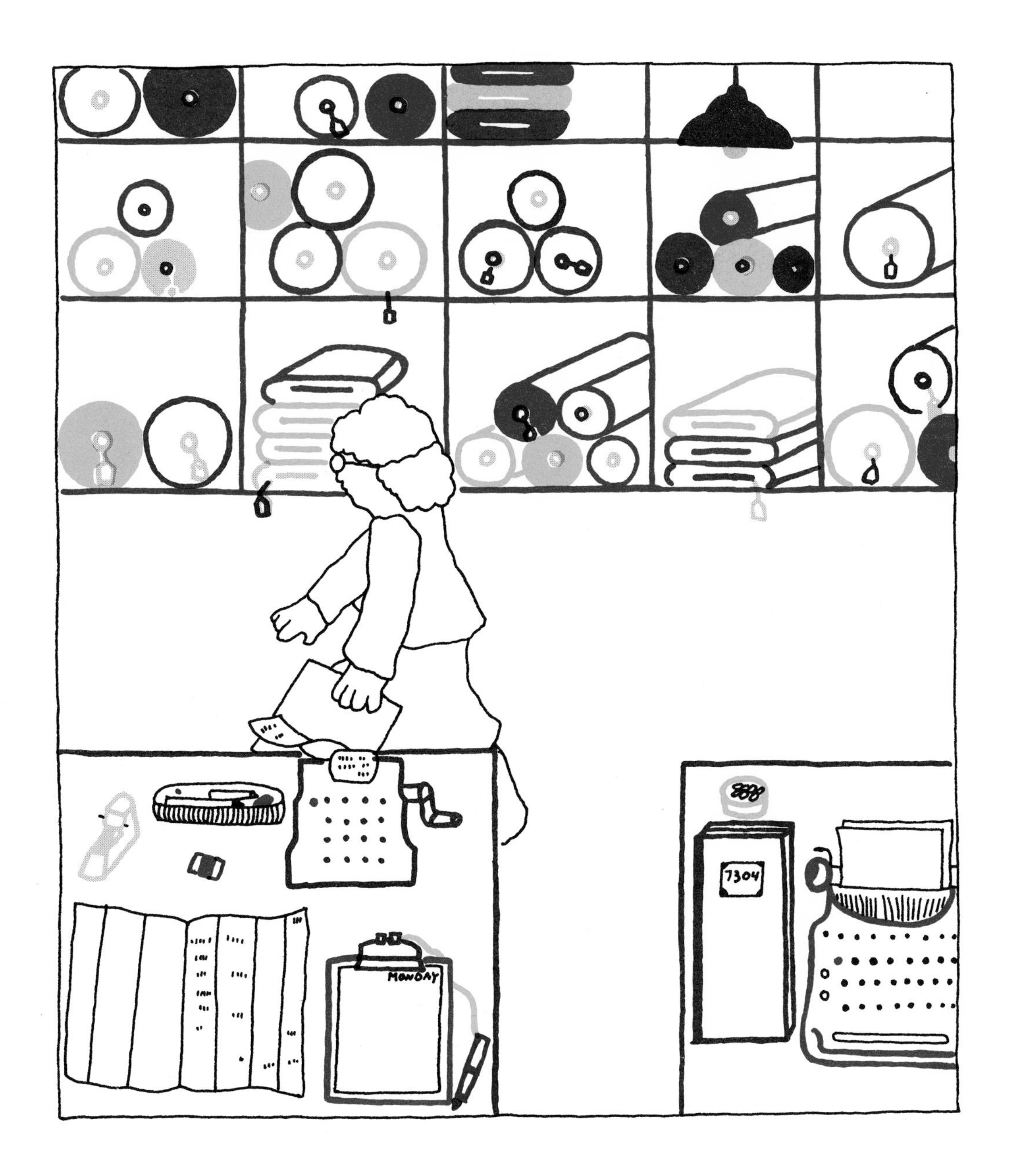

Fabric is numbered, ticketed, and stored in bins.

With a pair of good scissors, samples are carefully cut out from each new pattern. Samples are made for salesmen and customers.

To the showrooms come designers, clothing manufact- urers and store buyers, to choose new styles of fancy-weave twills, seersuckers, satin stripes, or plain-weave checks.

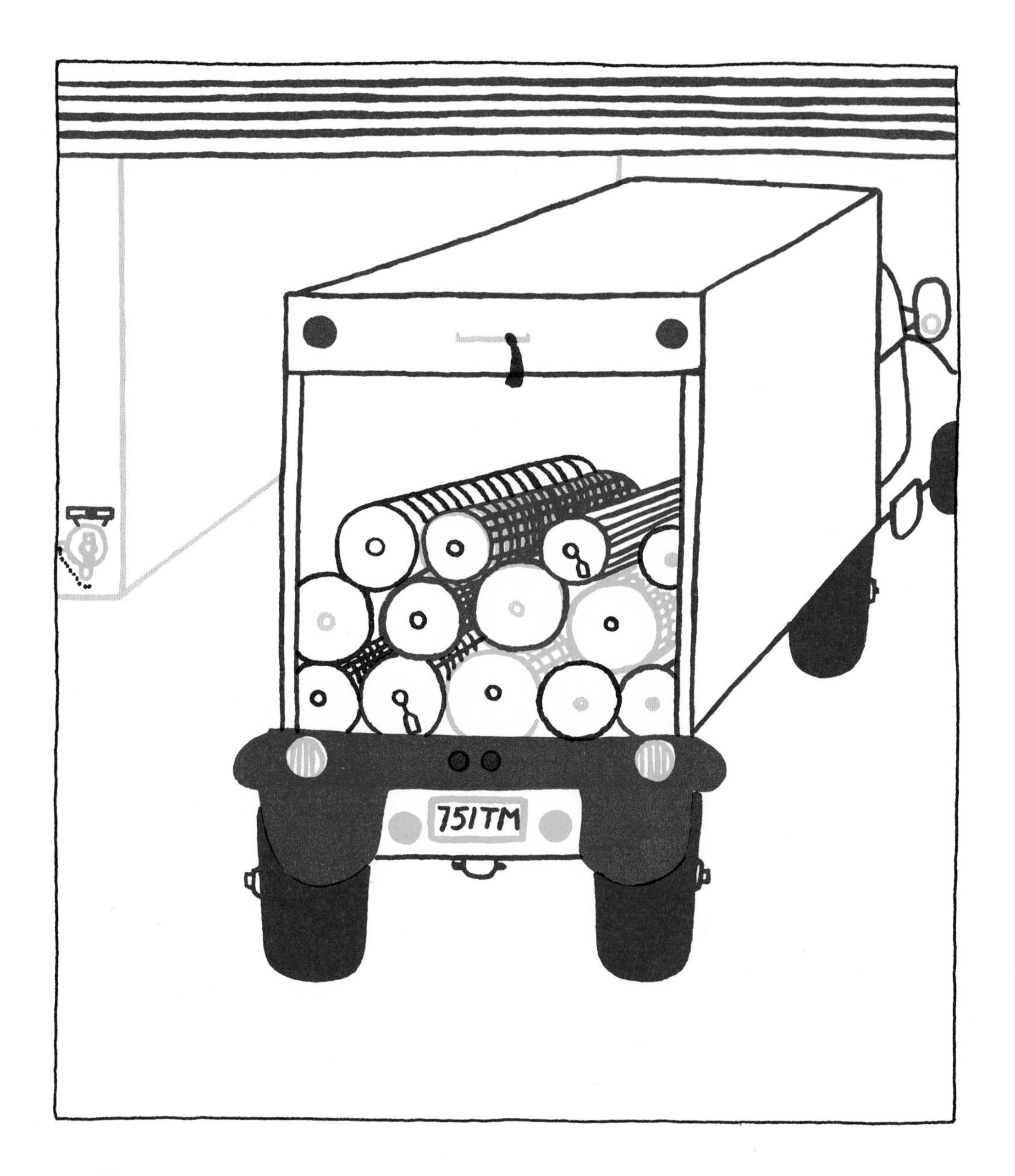

Rolls of fabrics are shipped to the customers by truck,

or wheeled down the street in push-carts.

Escalator up!

At the store, shoppers buy fabrics by the yard.

On Cotton Day at school, there is an exhibit to celebrate cotton-made articles: denim overalls, calico dresses, canvas sneakers, baseball caps, an umbrella, and a corduroy shirt.

Last-minute rehearsals before the fashion show: missing buttons are replaced, or a hem sewn. Cameras are clicking.

Cotton is King!

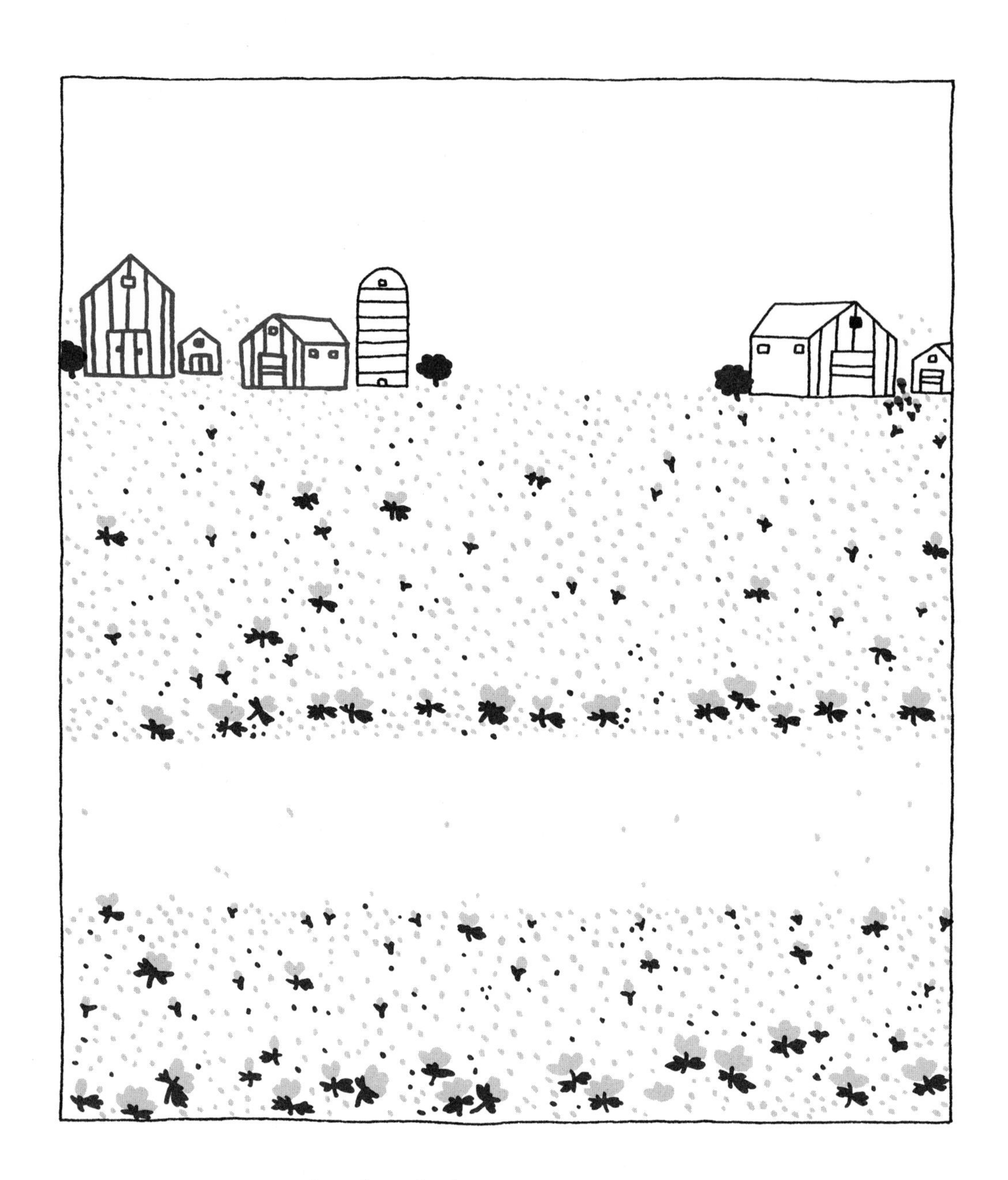

In the fields, next spring,

cotton will flower again!